NATURA NATURATA

Theory of Change for a National Pollinator Strategy in Portugal, Sílvia Castro, João Loureiro, Marten Schoonman, Martijn Thijssen, Anabela Nave, Andreia Penado, Adriana Galveias, Bárbara Oliveira, Carla Rego, Cláudia Fernandes, Elisabete Figueiredo, Eva Monteiro, Helena Ceia, Helena Freitas, Inês Santos, Joana Godinho, João T. Nunes, Luís Guilherme Sousa, Maria Luís Fino, Mário Boieiro, Mário Carmo, Ofélia Anjos, Sandra Antunes, Sofia Ferreira, Sónia Ferreira. Imprensa da Universidade de Coimbra;

ISBN: 978-989-26-2626-0

eISBN: 978-989-26-2627-7

DOI: https://doi.org/10.14195/978-989-26-2627-7

Institutional Partners:

PROMOTE POLLINATORS

polli. NET

Naturalis Biodiversity Center

THEORY OF CHANGE FOR A

NATIONAL POLLINATOR STRATEGY IN PORTUGAL

SÍLVIA CASTRO
JOÃO LOUREIRO
COORD.

EDITION

Imprensa da Universidade de Coimbra
Email: imprensa@uc.pt
URL: https://www.uc.pt/imprensa
Vendas online: https://livrariadaimprensa.uc.pt

EDITORIAL COORDINATION

Imprensa da Universidade de Coimbra

GRAPHIC CONCEPTION

Imprensa da Universidade de Coimbra

COVER IMAGE

João Loureiro

INFOGRAPHICS

Mickael Silva

PRINT BY

KDP

ISBN

978-989-26-2626-0

DIGITAL ISBN

978-989-26-2627-7

DOI

https://doi.org/10.14195/978-989-26-2627-7

LEGAL DEPOSIT

/25

SCIENCE COMMITTEE

Sílvia Castro
João Loureiro
Marten Schoonman
Martijn Thijssen
Anabela Nave
Andreia Penado
Adriana Galveias
Bárbara Oliveira
Carla Rego
Cláudia Fernandes
Elisabete Figueiredo
Eva Monteiro
Helena Ceia
Helena Freitas
Inês Santos
Joana Godinho
João T. Nunes
Luís Guilherme Sousa
Maria Luís Fino
Mário Boieiro
Mário Carmo
Ofélia Anjos
Sandra Antunes
Sofia Ferreira
Sónia Ferreira

TABLE OF CONTENTS

SUMMARY 9

INTRODUCTION International and national framework 11

THEORY OF CHANGE Conceptual framework 15

THE ToC PROCESS IN PORTUGAL Team and methodology 16

PORTUGUESE POLLINATOR ToC 19

CONTEXT STATEMENT 19

IMPACT STATEMENT 24

ACTORS Identified in Portugal 25

FACTORS Identified in Portugal 28

THEORY OF CHANGE DIAGRAM 31

Narrative construction 31

Reading the narrative 34

Contact 35

THEORY OF CHANGE NARRATIVE 37

THEME 1 - POLICY & FUNDING 37

PATHWAY *Policy & funding* 37

THEME 2 - RESEARCH & KNOWLEDGE 40

PATHWAY *Knowledge gaps and research* 41

PATHWAY *Monitoring* 42

THEME 3 - PRACTICES & MANAGEMENT 43

PATHWAY *Farming & forestry* 44

PATHWAY *Urban areas* 45
PATHWAY *Beekeeping* 46
THEME 4 - AWARENESS & ECOLITERACY 48
PATHWAY *Formal education* 48
PATHWAY *Non-formal and informal education* 49

FINAL OUTCOMES 50

FINAL COMMENTS 53

ACKNOWLEDGEMENTS 55

ABBREVIATIONS 57

GLOSSARY 59

BIBLIOGRAPHIC REFERENCES 63

APPENDIX 1 67

SUMMARY

BACKGROUND The Theory of Change (ToC) workshops supported by Promote Pollinators Alliance aim to assist the development of pollinator strategies by building a cause-effect roadmap on how specific actions can trigger intended outcomes, and how these subsequently contribute to the desired impact, i.e., pollinator conservation. This initiative provides an opportunity for members lacking a national governmental initiative focused on pollinators, such as Portugal, to develop the basis for a future Pollinator Action Plan.

METHODOLOGY The Portuguese Pollinator ToC engaged 24 participants and 2 international consultants, representing 23 organizations, in a series of participatory online meetings from April 2022 to July 2023. The meetings included 3 workshops with Promote Pollinators and Naturalis Biodiversity Center consultants, 15 intermediate meetings of the Portuguese team to develop the conceptual diagram, and 5 meetings to build the narrative presented here.

RESULTS During the development of the Portuguese Pollinator ToC, four main themes were identified, each with its pathway but intimately connected: policy & funding, research & knowledge, practices & management and awareness & ecoliteracy. Altogether, 92 actions were identified, including 54 activities and 38 outcomes. All pathways contributed to the following impact statement identified collectively: pollinators and the ecosystem services they provide are promoted across all landscapes (including natural, forest, agricultural and urban areas). The process was presented at the COP15 UN Biodiversity Conference, at the side event organized by the Promote Pollinators Alliance on pollination protection.

CONCLUSIONS The stepwise and co-construction ToC process enabled us to structure the (scientific) knowledge, opinions and ideas and to establish the bases for a future, impactful strategic plan. The process also enabled us to strengthen interpersonal relationships towards the construction of a consensual, jointly coordinated action plan that focuses on desired large-scale outcomes to effectively conserve pollinators.

INTRODUCTION
INTERNATIONAL AND NATIONAL FRAMEWORK

The **essential role of pollinators** was recognized by governmental entities, when the **Intergovernmental Science-Policy Platform on Biodiversity and Ecosystem Services** produced the assessment report on pollinators, pollination and food production (IPBES, 2016) and when the **Convention on Biological Diversity** (CBD) established the **International Pollinators Initiative** (IPI, 2016), coordinated by FAO, to protect pollinators and promote the sustainability of pollination services. The 2018-2030 action plan is aligned with the **Strategic Plan for Biodiversity**, the **Vision for Biodiversity 2050** and the **Sustainable Development Goals** (Agenda 2030). Additionally, the European Commission established the **European Pollinator Initiative** (2018-2020, EU Pollinator Initiative) and identified three priorities: improving knowledge about pollinator decline, its causes and consequences; fighting the causes of decline of this functional group; and raising awareness, involving society and promoting collaboration among stakeholders. The initiative defined **strategic objectives** and a set of **actions to be implemented by the EU and its Member States** to address pollinator decline in the EU and contribute to global conservation efforts. More recently, the EU Parliament adopted the **EU Nature Restoration Law** with key targets for ecosystem restoration and food security safeguard, which is currently being discussed to shape the legislation framework. In particular, **article 8** sets **important goals for pollinators**, namely the obligation to reverse pollinator decline and to achieve an increasing trend of pollinator populations until satisfactory levels are reached.

In Portugal, while a national governmental initiative specifically focused on pollinators and pollination is still lacking, the importance of these organisms is recognized in the **National Strategy for the Conservation of Nature and Biodiversity 2030**, in particular in Axis 2.1 "Promoting the mapping and assessment of the condition of ecosystems and improving their capacity to provide services relevant to well-being", which defines the creation of a **collaborative network at the national level** for the conservation of pollinators. Thus, in May 2021, **polli.NET – Collaborative network for the assessment, conservation and enhancement of pollinators and pollination** was created. Additionally, the importance of pollinators has been discussed in the Assembly of the Republic, which approved the **Resolution of the Assembly of the Republic n.º 336/2021**, of December 21st, 2021, and listed recommendations to the Government for measures regarding the preservation and conservation of pollinators and their habitats in the national territory.

The polli.NET Collaborative Network aims to promote pollinator conservation and pollination services sustainability through the development of a **concerted initiative on a national scale** that promotes **collaborative work between all stakeholders** and stimulates the safeguarding of the natural and functional heritage of high relevance to social well-being. The main objective of polli.NET is to **develop an action plan** for pollinators at the national level and to promote the implementation of measures proposed therein, involving and holding responsible all stakeholders.

In 2022, the **Promote Pollinator Alliance**, in collaboration with the **Naturalis Biodiversity Center**, offered a joint organization of a series of workshops based on the **'Theory of Change' method** with the mission to '*share, in an open and transparent manner, experience and lessons learnt in developing, facilitating and implementing pollinator strategies*'. A call was launched for the members of the Promote Pollinators Alliance, particularly for **countries lacking a Pollinator Action Plan**. The Portuguese national contact points applied and were among the countries selected. This initiative provided a unique opportunity to discuss the current stage, threats and actions needed for pollinators in Portugal, in an **open and friendly environment** where everyone **shared their experiences, knowledge, judgement and ideas** toward the common goal of pollinator promotion and conservation. This enabled us to **co-create the basis** for the development of a future National Action Plan for pollinators.

THEORY OF CHANGE
CONCEPTUAL FRAMEWORK

A **Theory of Change (ToC) is a methodology** to create an understanding and overview of how changes can be effectively induced toward a main goal. This method produces a **roadmap** (in the form of a diagram), that explains the **cause-effect logic of activities and a pathway of outcomes**. In particular, the diagram shows how specific activities (i.e., what is done by certain actors) lead to various, linked intended outcomes (i.e., trigger other stakeholders' behaviour) and how these then contribute to the desired impact (in this case, pollinator conservation). The ToC can be updated over time as the context, data and knowledge changes. The **diagram** is an excellent tool to show (some of) the pathways that can be followed to achieve the envisaged goals.

A Theory of Change is **developed using a participatory methodology** that gathers the knowledge and actors to arrive at a shared understanding of a **common main goal**. The ToC is shaped by the knowledge, experience, and beliefs of the people involved in its creation. Furthermore, participation fosters understanding of the intended results, who can do what and how collaboration play a role. Thus, it is very important to involve participants from different organizations to guarantee the inclusion of multiple perspectives and representatives of all stakeholders.

Added values of a ToC include:

- promote a **collaborative construction** – it is the basis for discussion in which the team comes to a solution together;

- demonstrate and promote a **collective understanding** of the **cause-effect changes**, i.e., what needs to be done by whom (activities) to achieve the intended outcomes;
- identify **actions or (preferably) outcomes** to be monitored to assess the impact of the activities towards the common goal.

Designing a ToC results in a **shared understanding** of the stakeholders involved, as well as their potential role in the cause-effect roadmap needed to achieve the desired impact.

THE ToC PROCESS IN PORTUGAL Team and methodology

The ToC process in Portugal involved a series of three online workshops with the consultants from the Promote Pollinators Alliance and the Naturalis Biodiversity Center, and a Portuguese team, to support the development of the Portuguese Pollinator ToC as a basis for a future national strategy. Fifteen extra meetings were promoted by the Portuguese team to develop the concepts and the ToC diagram, plus five meetings to develop the supporting narrative (section *Theory of Change Narrative*). The entire process was developed in an online format, enabling and promoting the participation of members from different parts of the country.

The main steps for this process were:

1. Identifying **country coordinators** and **team members**, including representation of different stakeholders;
2. Building a **context statement** based on the available knowledge of pollinators and pollination in Portugal;
3. Formulating the **impact statement** with a common main goal;
4. Identifying negative and positive **factors** impacting pollinators;
5. Identifying the **actors** involved;
6. Designing the ToC diagram: identifying **outcomes**, how these outcomes are **interlinked** in a logical action chain, **main pathways**

and **main activities**, and the underlying **assumptions** that are important to the operationalization of the ToC;
7. Writing a **narrative** to describe the Portuguese Pollinator ToC diagram.

The **team** included two country coordinators and 22 participants (**Appendix 1**) from 23 organizations representing various sectors that are relevant for a pollinator strategy, namely public administration of conservation and agriculture sectors, producers' associations, municipalities, crop protection industry, environmental consulting agencies, non-governmental organizations for environmental education and forest certification, academia, and education and science communication agencies. Country coordinators (Sílvia Castro and João Loureiro) were responsible for the coordination at the country level and were the contact persons with the Promote Pollinators Alliance and the Naturalis Biodiversity Center.

After gathering the team, first, the country coordinators, together with all participants, developed a **context statement** to describe the facts regarding pollinators and pollination in Portugal, ensuring that participants understood these facts and that all started from a common ground (section *Context statement*).

Second, the country coordinators ran a **preparatory survey** within the team identifying the dreams, worries, factors and actors, and transferred the information to a white-shared board. This **whiteboard was the working station** during all meetings. The information gathered in the survey served as a kickstart of **workshop 1** and the basis for drawing up the Theory of Change. In the first workshop, the team formulated, collectively, the **impact statement** (section *Impact statement*) and revised the main **actors and factors** impacting pollinators (section *Actors and Factors*), the latter were further developed in intermediate meetings of the Portuguese team. In **workshop 2**, the team started developing the diagram, defining **themes** based on the listed factors. The team was divided into breakout rooms devoted to specific themes and started formulating **outcomes**. The goal was to interlink outcomes in a logical order (grouped in pathways)

and identify concrete key **activities** needed to achieve the identified outcomes, starting from the impact statement and building the diagram from right to left in a sequence of cause-effect chain of actions. In **workshop 3**, the process was continued, by identifying the problems in the pathways and the assumptions needed in certain links. After developing the first complete version of the **cause-effect diagram**, the coordinators prepared a first draft of the narrative that was discussed in detail with the team through additional meetings, reaching the present document.

During the entire process, the participants identified a key problem in the communication between the different sectors - the terminology - being **difficult to have a consensus on the terminologies for territory**. This difficulty resulted not only from different fields of expertise, but also from different terminologies used in the available tools describing the territory. Thus, a **glossary of the terminology used** in the diagram and narrative was built (section *Glossary*), with a collective understanding and agreement of the definitions being used.

PORTUGUESE POLLINATOR ToC

CONTEXT STATEMENT

Portugal **recognizes pollination as a vital ecosystem service** for nature, agriculture, and human well-being, as recently stated by political parties in the Portuguese Parliament [1] and in the National Strategy for the Conservation of Nature and Biodiversity 2030 [2]. Additionally, Portugal mainland, together with the Azores and Madeira archipelagos, is an **important biodiversity hotspot**, harbouring a high diversity of wild pollinators and insect-pollinated plants (**Figure 1**). However, our **knowledge of pollinators and pollination is scarce**, and Portugal still **needs a national initiative** focused on pollinators and pollination.

POLLINATORS. While the information is systematized for the Azores and Madeira archipelagos in the form of checklists [3,4] and Azorean Biodiversity Portal [5], for the mainland, the available data is scarce and scattered. Portugal currently has a good knowledge of butterflies (133 species) [6]. In contrast, for bees, the first comprehensive lists are recent (from 2018 and 2022) [7,8], and the increased number of recent species descriptions and new species records [9-11] raise the number to over 740 species for mainland Portugal, a number that is likely to increase with further studies. Hoverflies are also poorly studied, counting over 200 species [12], with other pollinator groups being even less studied. Finally, only a few species are protected (Habitats Directive) [13].

Information about the trends of pollinator communities is still lacking, and monitoring schemes are yet in their infancy. However, through NGOs, Portugal has developed monitoring schemes based on citizen science to monitor Lepidoptera communities (butterflies implemented in 2019 [14], and moths implemented in 2021 [15]), engaging also monitoring activities in Classified areas. In 2022, Portugal also committed to testing the methodology of the EU Pollinator Monitoring Scheme within the SPRING project [16]. Furthermore, through polli.NET and PolinizAÇÃO project, a pilot project to test and propose a monitoring scheme is envisaged.

POLLINATION. There are also few studies available on pollinator communities of wild plants and their pollination efficiency. Following current estimates of pollinator dependence [17], over 78% of the flowering Portuguese flora may rely on pollinators to sustain their populations. Additionally, Portugal is characterized by a rich mosaic of land uses and diverse agroecosystems [18]. Over 57% of our crops depend on pollination by insects, with an increase in insect-pollinated crops in the last decade and pollination representing over 25% of the economic revenue (over 800 M€ per year) [19]. Therefore, the loss of pollinators will have severe impacts on agricultural production (both quantitative and qualitatively), with substantial economic implications for the sustainability of several key crops in Portugal.

KEY PRESSURES. Changes in land use and land cover, such as the loss of natural and semi-natural areas, homogenization of the landscape for agriculture and forestry or due to the expansion of invasive plants, management practices, as well as the misuse of pesticides, increasing occurrence of rural fires, and climate change are among the main key pressures identified worldwide [20] and present in Portugal. Urban intensification, with the reduction and fragmentation of green areas, soil sealing, and pollution are also important concerns. In addition, the extensive use of managed pollinators and invasive insects may also impact wild pollinators. However, only a few detailed studies on how these drivers impact our native entomofauna are available.

MADEIRA AND AZORES ARCHIPELAGOS. Pollinator diversity in Madeira and Azores is low compared to the mainland but has several exclusive species (24 endemic pollinators) [21-25] and unique interactions in the Portuguese territory, like vertebrate pollination [26]. In the last decade, the taxonomy, distribution and ecology of pollinators became better known in both archipelagos, particularly for bees, hoverflies and butterflies, and for the first time, comprehensive studies assessed their associations with plants and habitats. A few academic initiatives targeted pollinators and their services, several municipalities engaged in pollinator-friendly practices, while schools and science centres started developing initiatives to value pollinators. Despite the severe and extensive destruction and disturbance of native habitats of both archipelagos, particularly in the Azores, they still preserve remnants where native species and natural interactions are dominant. Also, some native generalist species play key roles as flower visitors in human-made environments [27].

Figure 1. Elements that can be found in Portuguese landscapes, including a summary for pollinators, pollination, key pressures and past and ongoing actions (for details see description in the text). Illustration by Margarida Moreira.

A world in which **pollinators and the ecosystem services they provide are promoted across all landscapes***

*Landscapes including natural, forest, agricultural and urban areas

This impact statement was the **starting point** for constructing the Portuguese Pollinator ToC diagram, building backwards the outcomes and actions needed to contribute to this impact.

ACTORS Identified in Portugal

The team has identified several **leading and secondary actors** in the **Portuguese context** and organized them according to their institutional typology, range of action and/or territorial coverage. These actors are listed below:

1) International organizations

- European Union
- International organizations for biodiversity conservation (e.g., IPBES – Intergovernmental Science-Policy Platform on Biodiversity and Ecosystem Services, FAO – Food and Agriculture Organization, CBD – Convention on Biological Diversity, Promote Pollinators Alliance)
- International Regulatory Bodies (e.g., EFSA – European Food Security Authority)

2) Governmental institutions at central and regional bodies (in the document referred as 'national governmental authorities' and referring to all governmental levels, from national to local level)

- Ministry of Environment and Energy
- Ministry of Agriculture and Fisheries
- Ministry of Education, Science and Innovation and its agencies (e.g., FCT – Foundation for Science and Technology)

- Ministry of Education
- Intermunicipal Communities and Municipalities (e.g., CIMs – Comunidades Intermunicipais)
- Commissions for the Coordination and Regional Development (CCDRs – Comissões de Coordenação e Desenvolvimento Regional)
- Secretaria Regional de Ambiente, Recursos Naturais e Alterações Climáticas, Região Autónoma da Madeira
- Secretaria Regional do Ambiente e Alterações Climáticas e Secretaria Regional da Agricultura e Desenvolvimento Rural, Região Autónoma dos Açores
- Municipalities

3) Governmental institutions dealing with biodiversity conservation and research

- ICNF – Instituto da Conservação da Natureza e das Florestas
- DGAV – Direção-Geral da Alimentação e Veterinária
- INIAV – Instituto Nacional de Investigação Agrária e Veterinária
- COTHN – Centro Operativo e Tecnológico Hortofrutícola
- IFCN – Instituto das Florestas e Conservação da Natureza, Região Autónoma da Madeira
- DRRF – Direção Regional de Recursos Florestais, Região Autónoma dos Açores

4) Scientific institutions from academy

- Research Institutions (e.g., Universities, Polytechnic Institutes)
- Collaborative Laboratories and Competence Centres (e.g., Centro de competências da Apicultura e Biodiversidade)
- R&D centres
- Natural History and Science Museums (e.g., National Natural History and Science Museum – MUHNAC, Science Museum of the University of Coimbra, Natural History and Science Museum of the University of Porto – MHNC-UP)
- Biological data platforms (e.g., Biodiversity4all)

5) Producer's organizations

- Farmers and their organizations (e.g., CAP - Confederação dos Agricultores de Portugal)
- Beekeepers and their organizations (e.g., FNAP, Fenapícola)
- Forest Owners and their organizations

6) Industry

- Plant protection industry (e.g., Syngenta)
- Environmental consultancy entities (e.g., NBI - Natural Business Intelligence)

7) Non-Governmental institutions (NGOs)

- NGOs (e.g., Tagis – Butterfly Conservation Centre of Portugal, Quercus, Zero, FSC – Forest Stewardship Council, Portugal, IUCN – International Union for Conservation of Nature, WWF – World Wildlife Fund)
- Societies (e.g., Portuguese Society of Entomology – SPEN, Portuguese Ecological Society – SPECO; Portuguese Botanical Society)

8) Formal and informal education institutions

- Science and Technology Education Centres (e.g., Ciência Viva)
- Schools (e.g., educational communities from pre-school, basic and secondary education schools)
- Elderly community centres

9) Dissemination institutions

- Media, journalists, and science communicators

10) Citizens

- Citizens communities
- polli.NET Collaborative Network
- Farmers and forest leaders

11) Trustees

It is important to highlight that, although linked with pollinators and pollination, the **authority or power of the stakeholders** listed above to affect the course of pollinator conservation activities and make decisions **varies greatly**, as well as their **level of awareness, concern and engagement** to develop the proposed actions.

Additionally, it is important to note that, although trustees were identified as actors, in Portugal, the involvement of these actors in nature conservation is not a common practice.

FACTORS Identified in Portugal

The team identified the **positive and negative factors impacting pollinators in Portugal** and organized them into themes according to their content. Below, we present the main themes and (some) related factors:

- **Policy and funding** – include the lack of directed funds for pollinator conservation, lack of a strategy to identify gaps, priorities and measures for pollinator conservation, lack of public policies directed to pollinators, lack of articulation among policies and policymakers at European, national and regional scales, bureaucratic impedance for implementing actions, lack of political concern about the importance of pollinators and inefficient knowledge transfer between science and policymakers.
  Yet, the team also identified an increased receptivity for pollinator conservation (e.g., Resolution of the Portuguese Parliament no. 336/2021 for pollinator conservation, establishment of a national network of stakeholders for pollinator conservation – polli.NET), increased engagement of several stakeholders and development of EU Pollinator Initiative frameworks for pollinators conservation.

- **Research and knowledge** – include the lack of data about diversity, current status and trends at local and national scales, lack of basic knowledge about the biology and ecology of pollinator species, lack of studies on the magnitude of major threats to pollinators in Portugal, lack of tailored-made solutions and practices for our territory, including, clear guidelines and capacitation for specific stakeholders, lack of robust methods for testing long term plant protection products (PPP) impacts (currently using few model species only), lack of entomologists and taxonomists due to long term disinvestment in advanced taxonomic training, and lack of tools for economic valuation of pollination services for stakeholders.

Yet, investment is being made in the first Invertebrates Red Book including key pollinator groups, research towards innovative and less risky PPPs, development of new agriculture and forest technologies towards more sustainable agricultural practices, and ongoing monitoring schemes including BMS-PT – Portuguese Butterfly Monitoring Scheme, Nocturnal Butterfly Station Network (REBN – Rede de Estações de Borboletas Nocturnas), and participation in SPRING project testing the EU-PoMS – European Pollinator Monitoring Scheme.

- **Practices and management** in agricultural, forest, and urban and peri-urban landscapes, and beekeeping – include the use of non-friendly pollinator practices, insufficient ecological infrastructures in some agricultural and forestry areas, inappropriate number and management of green urban areas in most municipalities, inappropriate management of road verges, and lack of native flora commercially available for gardening. Key pressures were also identified, including PPPs and their misuse, land use and land cover changes (e.g., landscape homogenization) and low landscape connectivity, pollution (e.g., air, light), climate change, increase frequency of rural fires, invasive animal and plant species, pathogens, predators and parasites, and ecosystem collapse.

Positive factors are the occurrence of traditional mosaic landscapes in certain regions and the floristic and entomologic richness of the territory, as well as the beginning of the development of ecological restoration in some areas, appropriate management not only by some agricultural and forest managers, but also of green urban areas in a few municipalities, and reduction in the use of PPPs. Also, while the impact of managed pollinators on wild pollinators remains unclear, there is a potential for optimizing beekeeping practices, promoting bee health.

- **Awareness and ecoliteracy** – include a global unawareness of pollinators' function, behaviour, diversity and importance, unawareness of economic impacts of pollinators in agriculture, unacceptance of pollinator-friendly actions by citizens (e.g., in urban green areas), fear of bees and other insects and inefficient knowledge transfer between science and citizens.
  While there is general illiteracy about pollinators and their importance, there is also increased receptivity for pollinator conservation and increased awareness by stakeholders from different sectors (e.g., farmers, municipalities and citizens).

The main **themes** identified here **originated the main pathways** of the Portuguese Pollinator ToC presented in this document.

THEORY OF CHANGE DIAGRAM

In Figure 2, we present the first version of the **Portuguese Pollinator ToC diagram** produced at the Promote Pollinator Alliance and the Naturalis Biodiversity Center workshops. Four main themes were identified, each with its pathway but intimately linked with each other, as follows: **policy & funding** coded in purple, **research & knowledge** coded in green, **practices & management** coded in yellow and **awareness & ecoliteracy** coded in blue.

The diagram was built around the **four themes** by identifying **outcomes** and **activities** and establishing a cause-effect relationship between them to achieve the **impact statement**. Through the following link, a digital version of the complete diagram can be open in a readable format: LINK

NARRATIVE CONSTRUCTION

This document translates the diagram into a narrative that describes how the relationships between the **activities** (coloured boxes) and **outcomes** (white boxes), via cause-effect relationships (indicated by the arrows), are built and ultimately lead to the achievement of the main goal (on the right), i.e., the **impact statement**. All the activities and outcomes are numbered to facilitate narrative and diagram reading.

The Portuguese Pollinator ToC is based on **four themes** tightly interconnected. The first theme is pivotal for the development of all actions, and the remaining three themes are identified as cornerstones for the development of a future action plan, as follows:

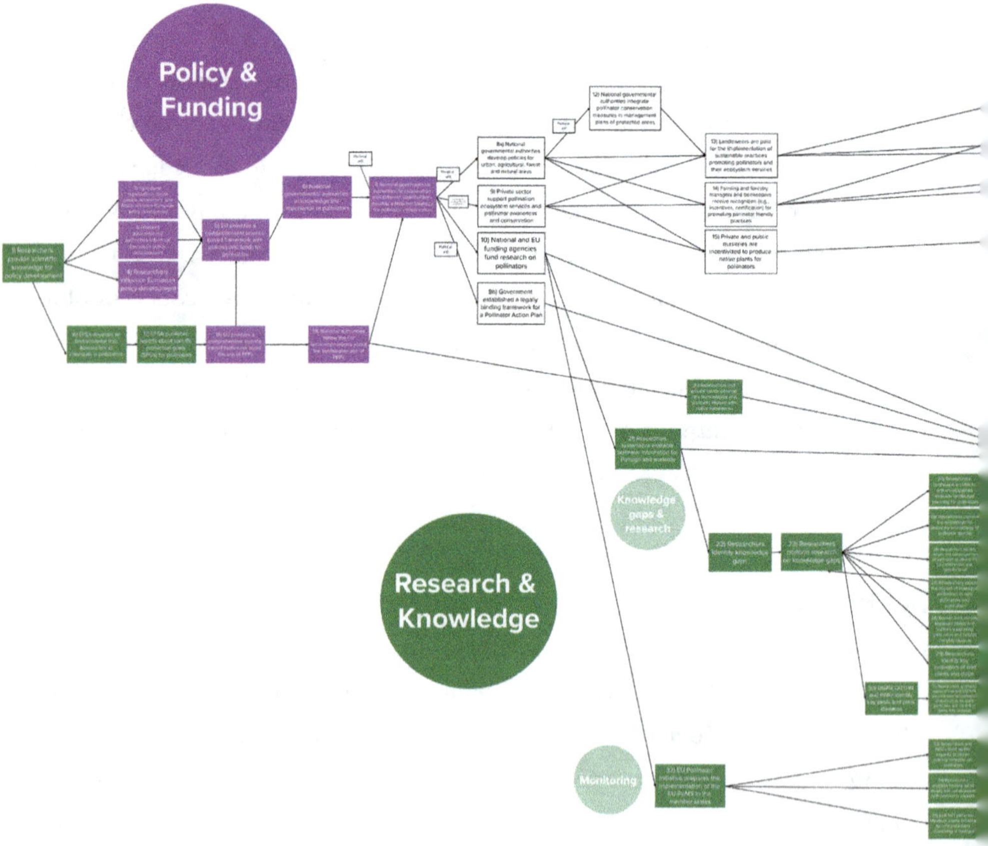

Figure 2. Diagram produced in the Portuguese Pollinator ToC identifying four main themes: policy & funding, research & knowledge, practices & management and awareness & ecoliteracy and how they are related (LINK).

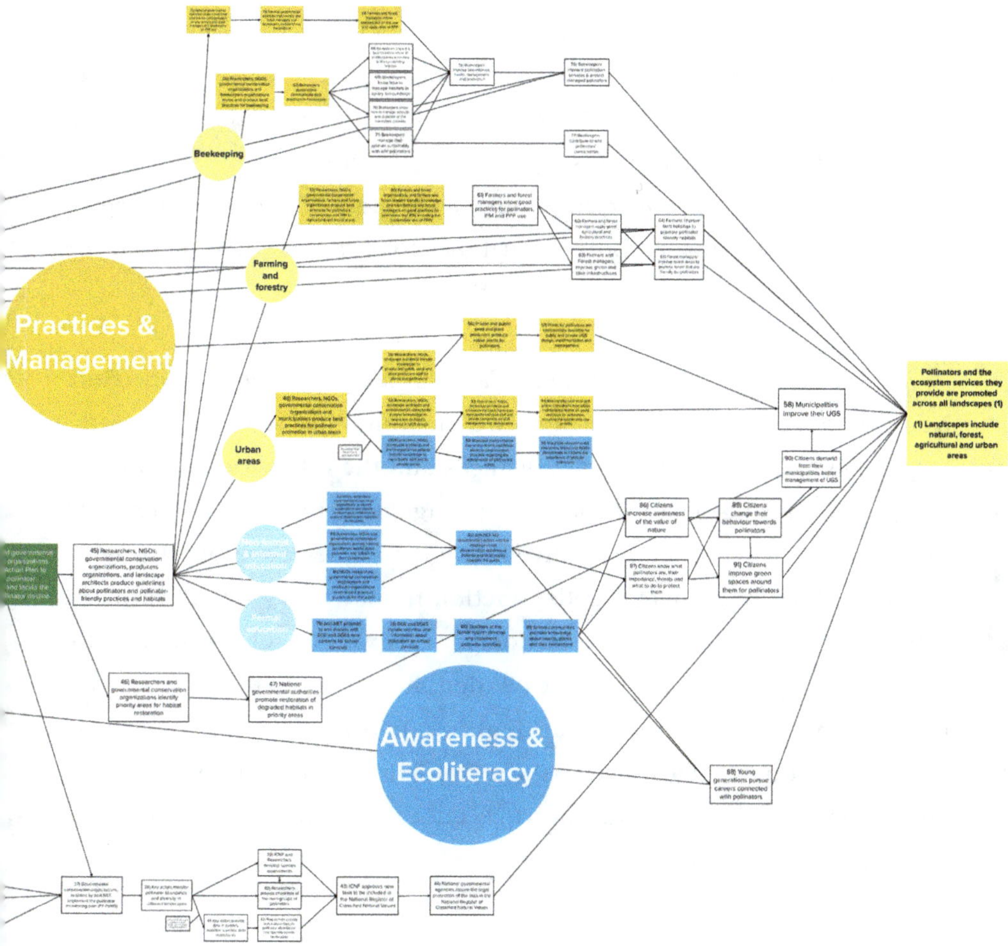
Beekeeping
Farming and forestry
Practices & Management
Urban areas
Awareness & Ecoliteracy
61) Farmers and forest managers know good practices for pollinators, IPM and PPP use
48) Researchers, NGOs, governmental conservation organizations and municipalities produce best practices for pollinator promotion in urban areas
58) Municipalities improve their UGS
Pollinators and the ecosystem services they provide are promoted across all landscapes (1)
(1) Landscapes include natural, forest, agricultural and urban areas
90) Citizens demand from their municipalities better management of UGS
86) Citizens increase awareness of the value of nature
89) Citizens change their behaviour towards pollinators
87) Citizens know what pollinators are, their importance, threats and what to do to protect them
91) Citizens improve green spaces around them for pollinators
45) Researchers, NGOs, governmental conservation organizations, producers organizations, and landscape architects produce guidelines about pollinators and pollinator-friendly practices and habitats
46) Researchers and governmental conservation organizations identify priority areas for habitat restoration
47) National governmental authorities promote restoration of degraded habitats in priority areas
88) Young generations pursue careers connected with pollinators

1. **Policy & funding:** raise political awareness, redirect funds and regulate actions towards the conservation of pollinators and their habitats.
2. **Research & knowledge:** promote and increase knowledge about pollinators and their habitats at the national level, develop monitoring programs, identify threats and develop solutions.
3. **Practices & management:** apply knowledge and act, developing concrete actions for the conservation of pollinators and their habitats across all landscapes of our territory.
4. **Awareness & ecoliteracy:** raise awareness of the importance of pollinators and mobilize and hold civil society accountable for promoting pollinator conservation.

Within each theme pathway, several **sub-pathways were identified**. Namely, for research & knowledge, we identified two distinct sub-pathways, one linked with research advances and the other focused on pollinator monitoring. For practices & management theme, we identified different key actors directly involved with the implementation of actions in the territory and having distinct action frameworks, namely farmers, foresters, urban managers and beekeepers, as well as related stakeholders. Thus, this theme was further divided in four sub-pathways. Finally, for awareness & ecoliteracy we identified two main ways to work on the citizen's knowledge, namely through the formal educational systems and through non-formal and informal education. Thus, this theme was divided into two sub-pathways.

READING THE NARRATIVE

Numbers in parenthesis **(#)** refer to specific activities or outcomes and can be located in the ToC diagram.

Bold sentences after a title are short introductions to the topic.

Additional **bold** sentences highlight concluding remarks.

Bold terms are intended to emphasize the important aspect and facilitate reading.

Sentences in *italics* are notes on the development of the cause-effect process.

Sentences where "assumption" is underlined indicate a causal assumption. This assumption explains why we think that outcome B is a logical consequence of outcome A.

The colour coding of the boxes follows the main themes identified above: **policy & funding** is coded in purple, **research & knowledge** is coded in green, practices & management is coded in yellow and **awareness & ecoliteracy** is coded in blue. Sub-pathways are represented in the diagram as circles in light shades corresponding to their respective themes. Note that **colour boxes are starting points, i.e., activities** that are necessary to achieve the underlying **outcomes (white boxes)**.

CONTACT

The Portuguese Pollinator ToC is a product of and for all the stakeholders linked directly and indirectly with pollinators and pollination in Portugal. Also, this exercise is not static, and the diagram will be continuously updated – note that the current version of ToC was made with the knowledge, expectations, and reality of a particular group of participants in a given period of time that can (and must) be adapted to new scenarios.

If you have questions, comments, or additions to the ToC, we would like to hear from you. For contacts, please use the following e-mail: pollinet@pollinet.pt.

THEORY OF CHANGE NARRATIVE

THEME 1 - POLICY & FUNDING

The activities and outcomes of the Theory of Change related to **policy & funding** are organized in one pathway (**Figure 2**) corresponding to the goal:

Raise political awareness, redirect funds and regulate actions towards the conservation of pollinators and their habitats.

INTRODUCTION *Policy & funding*

Legally binding guidelines, both at the national and European levels, are needed to implement and monitor actions towards pollinator conservation. Funding is crucial to trigger action, to build our knowledge of pollinators, and to foster action implementation by all actors involved. Lobbying politicians by disseminating knowledge and raising awareness for pollinator conservation is needed.

PATHWAY *Policy & funding*

ACTIVITIES: Policy measures related to pollinator conservation need to be **based on scientific knowledge**. Then, different actors will **lobby in favour of pollinators**, namely, (1) researchers need to provide scientific knowledge so that (2) agricultural organizations, forest owners, landowners and NGOs, as well as (3) national governmental authorities are better prepared and capacitated to influence European policy development. Also, (4) researchers need to be involved and influence European policy development as consultants. These lobbies may press the (5) EU to provide a comprehensive **science-based framework with policies and funds for pollinator monitoring and conservation**.

If this (5) science-based framework is defined and under the assumption that there is political will, (6) national governmental authorities will have to acknowledge the importance of pollinators and will have guidelines [including on (18) PPPs sustainable use], as well as funding to (7) develop a National Strategy for pollinator conservation, in cooperation with different stakeholders. *A National Strategy is a general plan based on strategic thinking and strategic planning to achieve long-term overall goals under conditions of uncertainty. It is important to note that a National Pollinator Strategy must include an Action Plan. Also, it is necessary to work towards a closer cooperation between governmental institutions from different sectors. Finally, it is important to identify actions to promote political will, here defined as the commitment of policy makers to the discussion and development of solutions in favour of pollinators.*

OUTCOMES: If a (7) National Strategy for pollinator conservation is created, then a framework for policy measures will be created. Under the assumption that there is political will, (8a) national governmental authorities will develop policies for urban, agricultural, forest and natural areas, and (8b) establish a legally binding framework for a National Pollinator Action Plan. **These outcomes will trigger action and foster the implementation of measures by all actors involved.**

Also, under the assumption of an overall environmental responsibility perception and/or obligation through legislation, the (7) National Strategy for Pollinator Conservation may lead to (9) the private sector supporting pollination ecosystem services and pollinator awareness and conservation.

The above outcomes are expected to trigger: (12) national governmental authorities to integrate pollinator conservation measures in the management plans of protected areas (once again under the assumption of political will); (13) payments to landowners for the implementation of sustainable practices promoting pollinators and their ecosystem services; (14) farming and forestry managers and beekeepers to receive recognition (e.g., incentives, certification) for promoting pollinator-friendly practices; and (15) private and public nurseries to be incentivized to produce native plants for pollinators. These activities will result in several outputs directly linked with our main goal, i.e., the promotion of pollinators and their ecosystem services across all landscapes, either directly by (64) the improvement of crop and non-crop areas to promote pollinator-friendly habitats by farmers, (65) the improvement of forest areas to promote forests that are friendly for pollinators by forest managers, (76) the improvement of pollination services and protection of managed pollinators by beekeepers, or indirectly by (58) the improvement of urban green spaces (UGS) in pollinator-friendly ways by municipalities and citizens. *Again, it is important to identify activities to promote political will and connect knowledge with the interests and mission of the actors at governmental and non-governmental institutions.*

Finally, (6) the recognition of pollinators and their importance and (7) the development of a National Pollinator Strategy will drive (10) national and EU funding agencies to fund research on pollinators, feeding the development of the **research & knowledge** pathway.

ACTIVITIES: Additionally, there is an **important sub-pathway that supports the science-based framework devoted to PPPs actions and outcomes**. Namely, (1) if knowledge produced by researchers is made available for policy development, including PPPs impact on pollinator's health, then (16) EFSA will be pressed to develop an Environmental Risk Assessment of chemicals on pollinators, and (17) to define specific protection goals (SPGs) for pollinators. This will enable the (18) EU to provide a **comprehensive science-based**

framework for the use of PPPs, and (19) national authorities are expected to follow the EU recommendations about the sustainable use of these products. *It is important to note that the impact assessment of PPPs has been done mainly on honeybees and, more recently, Bombus terrestris and Osmia bicornis (all managed pollinators). Thus, it is recommended that the assessments are widened to representatives of other pollinator groups, acknowledging the difficulty to operationalize this recommendation.*

To address EU recommendations, (20) researchers and the private sector will develop new technologies and products aligned with policy frameworks, contributing to (11) the development of the Pollinator Action Plan. **The development of an Action Plan for Pollinators defines and prioritizes the actions and identifies the actors involved in the promotion and conservation of pollinators and pollination services.**

THEME 2 - RESEARCH & KNOWLEDGE

The activities and outcomes of the Theory of Change related to **research & knowledge** are organized in two sub-pathways (**knowledge gaps and research** and **monitoring**, **Figure 2**), corresponding to the goal:

Promote and increase knowledge about pollinators and their habitats at the national level, develop monitoring programs, identify threats and develop solutions.

INTRODUCTION *Research and knowledge*

The development and systematization of the scientific knowledge are an important basis for defining adequate actions for pollinators. Scientific knowledge has an added value when it is made available in an accessible manner to all the actors involved.

PATHWAY *Knowledge gaps and research*

ACTIVITIES: If (10) National and EU funding agencies fund research on pollinators' conservation, funds will be available for (21) researchers to systematize available pollinator information for Portugal (and worldwide if needed). This will enable (22) researchers to identify knowledge gaps and (23) perform research on these areas. These activities will trigger the (11) **development of the National Action Plan** to promote pollinator conservation and tackle the causes of pollinator decline.

Based on the **knowledge gaps already identified** in our territory, several research lines are recognized, namely: (24) research on landscape planning for pollinators, (25) improve knowledge of the taxonomy and biology of pollinator species, (26) identification of the causes and consequences of pollinator decline in Portugal (both at community and species level), (27) assess the impact of managed pollinators in wild pollinators and pollination, (28) identification of important plants and habitats supporting pollinators and habitat carrying capacity, and (29) identification of key pollinators of wild plants and crops. Also, (30) DGAV, COTHN and INIAV need to identify key pests and plant diseases that will enable (31) researchers, producer associations and COTHN to develop new technologies and practices for plant protection and pest and disease control. **This knowledge provides a more complete picture of the threats and opportunities to improve the conditions for pollinators, and helps all actors involved to act towards pollinators' promotion and conservation.** *Note that further gaps in our knowledge of pollinators can be identified, and research lines can (and must) be added in the future.*

OUTCOMES: If (11) an Action Plan is developed and (10) funds are available, (46) researchers and governmental conservation organizations will identify priority areas for habitat restoration and drive (47) the restoration of degraded habitats in priority areas, directly contributing to pollinator conservation.

At the same time, (45) researchers (e.g., entomologists, botanists, and phytosociologists), NGOs, governmental conservation organizations, producer's organizations and landscape architects will produce guidelines about pollinators and pollinator-friendly practices and habitats for different landscapes of the territory (agricultural, forest and urban areas) and for beekeeping. These guidelines have the potential to trigger activities and outcomes in important pathways of **practices & management** and **awareness & ecoliteracy** towards our main goal (see below).

The **Action Plan for pollinators will assume special relevance** if the (8b) government establishes a legally binding framework and (10) awards specific and adequate funds for its implementation.

PATHWAY *MONITORING*

ACTIVITIES: In the context of the (32) EU Pollinator Initiative support for the implementation of EU-PoMS in member states, and if (10) funds for pollinator conservation are available, (33) researchers and NGOs will build up the capacity of citizen science networks on pollinators, (34) researchers will promote training (including basic, intermediate and advanced training) and foster collaboration with taxonomic experts, and (35) polli.NET will pilot the Minimum Viable Scheme for wild pollinators monitoring in Portugal. Altogether, this will contribute to the (36) definition of the EU methodology for pollinator monitoring to be implemented by member states. **This provides the baseline for the development of a national monitoring program, in the future, legally binding for the EU member states.**

OUTCOMES: If (36) the EU defines the methodology for pollinator monitoring by member states and (11) an Action Plan is devised including a pollinator monitoring plan, (37) governmental conservation organizations assisted by polli.NET will implement the Portuguese

Pollinator Monitoring Scheme (PT-PoMS). Then, (38) key actors will monitor pollinator abundance and diversity in different landscapes. Under the assumption that public funds are provided, (41) monitoring data is expected to be publicly available in scientific repositories, enabling (42) researchers to analyse the abundance and diversity of pollinators across the territory, and assisting the (39) development of species assessments and the (40) compilation of updated checklists of the main pollinator groups. **Monitoring is crucial to establish a baseline and assess trends in pollinator populations across landscapes in Portugal, as well as to evaluate the impact of actions in favour of pollinators implemented in the territory.**

If the above-mentioned actions are accomplished, data will be available to support the (43) approval of new taxa to be included in a future National Register of Classified Natural Values by ICNF, legally binding in Portugal. (44) National governmental agencies will have to ensure not only the legal protection of these taxa but also the needed funds, directly contributing to the legal protection of these organisms. *Note that this process will be led by ICNF in straight collaboration with conservation authorities from the Azores and Madeira Autonomous Regions.* **The identification of endangered taxa is crucial for the definition of habitat restoration priorities and for supporting the development of environmental assessment reports.**

THEME 3 - PRACTICES & MANAGEMENT

The activities and outcomes of the Theory of Change related to **practices & management** are organized in three sub-pathways (**farming & forestry**, **urban areas** and **beekeeping**, **Figure 2**), corresponding to the goal:

Apply knowledge and act, **developing concrete actions** for the conservation of pollinators and their habitats across all landscapes of our territory.

INTRODUCTION *Practices & management*

Farmers and forest managers, beekeepers, governmental institutions, and citizens can make a difference by adopting pollinator-friendly measures within the scope of their activities. An equilibrium between nature conservation and farming and forestry development ensures a better living environment for pollinators. This equilibrium promotes diverse and abundant pollinator communities and, consequently, better pollination services and sustainable agricultural production. Additionally, gardening and landscaping sectors play an important role in providing pollinator-friendly solutions for public and private green areas that function as ecological corridors or stepping stones in the urban matrix. Altogether, we can work on multifunctional landscapes that serve for people and nature.

PATHWAY *Farming & forestry*

ACTIVITIES: Once (45) guidelines on pollinators and pollinator-friendly practices and habitats are produced, (59) researchers, NGOs, governmental conservation organizations, farmers and forest organizations may produce best practices for pollinator conservation and for integrated pest management (IPM) in agricultural and forest areas. Then, (60) farmers and forest organizations, and farmers and forest leaders will transfer this knowledge and train farmers and forest managers on good practices for pollinators and IPM, including the sustainable use of PPPs. *This pathway still needs to be further developed to discuss topics such as PPP use, intensive agricultural and forestry practices, and increased extension of continuous intensive agricultural and forestry areas.*

OUTCOMES: If (60) farmers and forest managers are trained on the best agricultural and forestry practices towards biodiversity promotion and conservation, including pollinators, then (61) they will know and

(62) apply them, (63) improving green and blue infrastructures, in a way where both (64) farms and (65) forest areas are improved to promote pollinator-friendly habitats. **This guarantees food and nesting resources to sustain viable populations of wild pollinators, directly contributing to pollinator promotion and conservation.**

PATHWAY *Urban areas*

ACTIVITIES: In parallel, once (45) guidelines on pollinators and pollinator-friendly practices are produced, (48) researchers, NGOs, governmental conservation organizations and municipalities will produce best practices for pollinator promotion in urban areas. The implementation of best practices in urban areas will benefit from the **awareness & ecoliteracy** pathway, where (49) researchers, NGOs, landscape architects and environmental consultants transfer this knowledge to municipality staff and to the private sector. Then, (50) municipal environmental awareness teams and NGOs will develop dissemination activities and materials regarding pollinator-friendly UGS management and (51) will inform citizens about their importance. *To produce change, knowledge transfer needs to acknowledge the costs and benefits of current practices and of pollinator-friendly practices.*

OUTCOMES: These actions, (86) will increase nature value awareness, namely in urban areas, and (89) will change citizens' behaviour towards pollinators and (90) a demand for better management of UGS. **These activities are crucial so that the measures implemented afterwards in urban areas are understood and well-received by citizens.**

ACTIVITIES: Also, if (52) researchers, NGOs, landscape architects and environmental consultants transfer the knowledge to landscape architects involved in the design of UGS and (53) train municipality technical staff and the private sector on UGS pollinator-friendly management and maintenance, then, (54) municipality technical staff and/or consultants will train UGS maintenance teams on good practices for pollinators, including the sustainable use of PPPs. *It is important to note that the chain of actors involved with UGS is very complex and involves both public and private actors that may be different depending on local administrations.*

Finally, (48) best practices in urban areas, together with (15) subsidies to produce native plants for pollinators, will also trigger a chain of action in private and public seed and plant producers. Thus, if (55) researchers, NGOs, and landscape architects transfer knowledge to private and public seed and plant producers on plants and pollinators, (56) they will invest in the production of native plants for pollinators, making them (57) commercially available for public and private UGS design, implementation and management.

OUTCOMES: Altogether, the (57) commercial availability of native plants for pollinators, the (54) UGS maintenance teams' knowledge about best practices for pollinators in these areas, and the (51) citizens' awareness of the need to create pollinator-friendly UGS will lead (58) municipalities and citizens to improve their UGS. **These outcomes improve urban landscape, providing ecological corridors and oases for the survival of pollinator communities in urban areas, contributing directly to pollinator promotion and conservation.**

PATHWAY *BEEKEEPING*

ACTIVITIES: Once again, if (45) guidelines about pollinator-friendly practices and habitats are produced, then (66) researchers, NGOs, governmental conservation organizations and beekeeper's organizations will be able to revise and produce best practices specifically for beekeeping. Then, (67) these best practices can be communicated to beekeepers by their organizations. This will enable beekeepers to (68) know the best locations where to install apiaries according to the surrounding habitats, (69) know how to manage habitats in the apiary surroundings, (70) know how to manage parasites and diseases of honeybee colonies, and (71) manage their apiaries sustainably with wild pollinators accounting with habitat carrying capacity and sustainable coexistence. *It is important to note that most beekeepers install their apiaries on private land and, thus, rely also on the best practices adopted by farmers and forest landowners and managers [outcomes (62) and (63)].*

Additionally, (45) guidelines on pollinator-friendly practices and habitats will trigger (72) national governmental agencies to create a functional platform about the use of PPPs for better communication between beekeepers and farmers and forest managers. Then, (73) national governmental agencies will train farmers and forest managers and beekeepers on how to use the platform, enabling (74) farmers and forest managers to inform beekeepers on the use of PPPs. *We would like to note that the legislation to inform about PPP use already exists, however, the communication is not effective. Thus, it is necessary to implement a tool to efficiently manage PPPs use among different actors.*

OUTCOMES: If (74) farmers and forest managers inform beekeepers on the use of PPPs, and beekeepers engage on (68-71) pollinator-friendly actions, (75) beekeepers will improve honeybee health, management and production and, at the same time, they will (76) improve pollination services and protect managed pollinators. **The implementation of best practices by beekeepers enables them to protect managed pollinators and the pollination services they provide.**

Also, if (71) beekeepers manage their apiaries sustainably with wild pollinators, for example, accounting for the habitat carrying capacity and competition between managed and wild pollinators, they will (77) contribute to wild pollinators' conservation. **This enables to attain an equilibrium between managed and wild pollinators, assuring beekeeping production and the survival of wild pollinator communities, as well as the ecosystem services they provide.**

THEME 4 - AWARENESS & ECOLITERACY

The activities and outcomes of the Theory of Change related to **awareness & ecoliteracy** are organized in two sub-pathways (**formal** and **non-formal & informal education**, **Figure 2**), corresponding to the goal:

Raise awareness of the importance of pollinators and **mobilize and hold civil society accountable** for promoting pollinator conservation.

INTRODUCTION *Awareness & ecoliteracy*

Raising awareness and changing behaviours through formal, non-formal and informal education is crucial to engage citizens and different stakeholders in pollinator conservation. It is a vehicle to help a variety of actors, in different learning scenarios, to be aware and take the appropriate pollinator-friendly actions. Knowledge transfer needs to acknowledge the costs and benefits of current practices and of pollinator-friendly practices to produce change. *Formal learning about pollinators is mainly promoted in learning institutions (e.g., schools, colleges, universities). Nevertheless, other learning activities can occur outside the formal educational system, including non-formal education, where organizations outside the established formal system develop educational activities towards pollinator conservation; and informal education, where citizens are engaged with activities without a learning purpose in mind.*

PATHWAY *Formal education*

ACTIVITIES: If (45) guidelines about pollinators and pollinator-friendly practices and habitats are produced, the (78) members of polli.NET will be able to propose and discuss with the Directorate General of Education (DGE) and the Directorate General for Higher Education (DGES) new contents for school curricula. Then, (79) DGE and DGES will include activities and information about pollinators in school curricula. This will allow (80) teachers in the formal education system to develop and implement pollinator activities that (81) promote knowledge about insects, plants and their interactions in educational communities. **As part of formal education, the value of pollinators and pollinator-friendly practices becomes an intrinsic component of citizens' knowledge, from an early age.**

OUTCOMES: Because both contribute to the same outcomes, formal education outcomes are presented below, together with non-formal and informal education outcomes.

PATHWAY *Non-formal and informal education*

ACTIVITIES: At the same time, the (45) guidelines might be used by different actors as grounds for dissemination activities, hands-on training and development of education materials. Thus, (82) NGOs, researchers, governmental conservation organizations, producer's organizations and science communicators will collaborate to produce dissemination materials about pollinators and pollination for the public; (83) researchers, NGOs, and governmental conservation organizations will provide training (at different levels) about pollinators and actions for their conservation; and (84) NGOs, researchers, governmental conservation organizations and producer's organizations will develop best practices guidelines directed to the public. Whenever possible, these activities should be supported by science communicators. The (85) dissemination of all these activities will be promoted at national and local scales towards the public by polli. NET, key dissemination actors (e.g., museums, science centres) and the media. **These activities disseminate the importance of pollinators and promote common knowledge among citizens.**

Additionally, within the **practices & management** pathway, there is an important link with **awareness & ecoliteracy**. This link is included in the sub-pathway of urban areas, as citizens' perception (negative or positive) regarding management practices implemented in UGS, is crucial for their acceptance. Thus, if (48) researchers, NGOs, governmental conservation organizations and municipalities produce best practices for pollinator promotion in urban areas, the implementation of best practices in urban areas will benefit from the **awareness & ecoliteracy** pathway, where (49) researchers, NGOs, landscape architects and environmental consultants transfer this knowledge to municipality staff and to the private sector. Then, (50) municipal environmental awareness teams and NGOs will develop dissemination activities and materials regarding pollinator-friendly urban green spaces (UGS) management and (51) will inform citizens about their importance. **These activities ensure understanding and receptivity of UGS pollinator-friendly management by citizens.**

OUTCOMES: The activities developed through formal, non-formal and informal education will contribute to increasing not only (86) citizens' awareness of nature's value, but also (87) knowledge of pollinators, their importance, threats and what to do to protect them. This ecoliteracy is expected to (89) change citizens' behaviour towards pollinators and (90) increase the demand for better management of UGS. These outcomes will lead to the (91) improvement of green spaces for pollinators. **This awareness mobilizes civil society for pollinator conservation, directly contributing to our main goal.**

The (81 & 85) engagement with pollinators through education will lead (88) young generations to pursue careers connected with pollinators and nature conservation, strengthening the long-term pathway of **research & knowledge** and fuelling the entire chain of actions and outcomes.

FINAL OUTCOMES

The results of the four pathways will lead to the following main outcomes:

- (44) National governmental agencies assure the legal protection of pollinators listed in the National Register of Classified Natural Values;
- (47) National governmental authorities promote restoration of degraded habitats in priority areas;
- (89 & 90) Citizens change their behaviour towards pollinators and demand for pollinator-friendly green spaces management;
- (91 & 58) Citizens, school and elderly communities, and municipalities improve green spaces around them for pollinators;
- (64 & 65) Farmers and forest managers improve farms and forest areas to promote pollinator-friendly habitats;
- (76 & 77) Beekeepers improve pollination services, protect managed pollinators and contribute to wild pollinator conservation.

Together, they contribute to the impact statement: **Pollinators and the services they provide are promoted across all landscapes** of our territory.

FINAL COMMENTS

The present document is a product developed by a group of participants, members of the polli.NET, in the ToC workshops for all stakeholders directly and indirectly involved with pollinators and pollination in Portugal. This exercise reflects the knowledge, expectations, and reality of the group of participants in a given period of time that can (and must) be improved and adapted to new scenarios. Thus, this version of the ToC pathway and narrative is not static, and the goal is to **continuously update and improve it** with the help of everyone interested in participating actively. The next update will be developed within the project PolinizAÇÃO by the end of 2025.

The goal of Promote Pollinators ToC workshops was to support and provide the foundation for the development of national action plans for pollinator conservation and triggered the project **PolinizAÇÃO** (2023-2025), recently **funded by *Fundo Ambiental* from the Ministry of Environment and Climate Action** (current Ministry of Environment and Energy). These actions aim at **developing a legally binding and continuously funded Action Plan for Pollinator Conservation in Portugal**. The factors, actors and numerous key activities are identified and are likely to trigger specific outcomes towards pollinator conservation. A close cooperation between all stakeholders, as well as closer cooperation between governmental institutions from different sectors, will enable us to successfully develop, establish and overseeing the Pollinator Action Plan.

Finally, this document also intends to **trigger action** combined with a common understanding of what is needed and who develops it. Action towards pollinator conservation can be triggered by different stakeholders at any moment of the pathway. By being made available publicly (including a translation to Portuguese and preparation of a shorter and visual version), the team hopes to inspire different stakeholders identified in the diagram to start developing initiatives towards pollinator conservation. In the meantime, the participation in Promote Pollinators ToC workshops enabled us to present the results at COP15, raising awareness of Portuguese authorities to channel funds for the development of the **Action Plan for Pollinator Conservation in Portugal – PolinizAÇÃO**.

ACKNOWLEDGEMENTS

We would like to deeply thank the Promote Pollinator Alliance for organizing the Theory of Change workshops and the Naturalis Biodiversity Center, providing the opportunity to countries without Pollinator action plans to bring the discussion to the table. In particular, we are thankful to Martijn Thijssen and Marten Schoonman for all the guidance, prompt availability, knowledge sharing and friendship throughout the entire process. We are also thankful to Paulo A. V. Borges and Anabela Belo for their valuable comments on the document, and to Carolina Caetano and Cândida Ramos for their assistance with the diagram translation and the final review. Finally, the coordinators would like to leave a public acknowledgement to everyone who voluntarily gave their time and energy to share their views and knowledge, contributing to an open and friendly discussion that culminates in the present document.

ABBREVIATIONS

CAP – Confederação dos Agricultores de Portugal

CBD – Convention on Biological Diversity

COTHN – Centro Operativo e Tecnológico Hortofrutícola (https://www.cothn.pt/)

DGAV – Direção Geral da Alimentação e Veterinária (https://www.dgav.pt/)

DGE – Directorate General for Education

DGES – Directorate General for Higher Education

EFSA – European Food Safety Authority (https://www.efsa.europa.eu/en)

EU-PoMS – European Pollinators Monitoring Scheme (https://publications.jrc.ec.europa.eu/repository/handle/JRC122225)

FAO – Food and Agriculture Organization of the United Nations

ICNF – Instituto da Conservação da Natureza e das Florestas (https://www.icnf.pt/)

INIAV – Instituto Nacional de Investigação Agrária e Veterinária, I.P. (https://www.iniav.pt/)

IPM – Integrated pest management

NGOs – Non-governmental organizations

polli.NET – Rede Colaborativa para a Avaliação, Conservação e Valorização dos Polinizadores e da Polinização (https://www.pollinet.pt/)

PPP – Plant Protection Products

PT-PoMS – Portuguese Pollinator Monitoring Scheme

SPGs – Specific Protection Goals

SPRING – Strengthening Pollinator Recovery through Indicators and monitorinG (https://www.ufz.de/spring-pollination/)

ToC – Theory of Change

UGS – urban green spaces

GLOSSARY

Agricultural land – refers to land devoted to agriculture, consisting of annual crops, permanent crops, agricultural fallows, protected agriculture (e.g., polytunnels and greenhouses) and plant nurseries [30].

Agriculture intensification – farming system where management practices aim to increase the yield per unit of farmland or pasture.

Agroecosystems – communities of plants and animals interacting with their physical and chemical environments that have been modified by people to produce food, fibres, fuel and other products for human consumption and processing.

Agroforestry areas – refers to areas with vertical intercropping of temporary crops and/or pastures (improved or spontaneous) and permanent crops of forest species.

Apiary – set of bee colonies in production conditions, including the settlement site itself and respective infrastructure, belonging to the same beekeeper.

Beehive – physical support to house a bee colony and its production consisting of a bottom board, brood nest, honey supers, top cover and removable frames to support the combs.

Beekeeper's organizations – collective bodies with a non-profit associative and/or cooperative base grouping beekeepers with the main goal of representing, defending and promoting their interests.

Biological invasions – when a species is introduced deliberately or inadvertently by humans and establishes populations far from its native home, maintains itself without human assistance, and spreads beyond the point of introduction.

Citizen science – scientific research developed in a collaborative framework between professional scientists and the public engaged in data collection and analysis.

Classified natural values – natural values that, due to their relevance for the conservation of nature and biodiversity, are subject to national and international legal protection.

Crops – plants that are grown for food; a plant or plant product that can be grown and harvested extensively for profit or subsistence (food, or other goods).

Ecosystem – a community of animals and plants interacting with one another and with their physical environment.

Ecosystem services – refer to benefits that people obtain, directly or indirectly, from ecosystems, including: goods produced or supplied by ecosystems; benefits obtained from the regulation of ecosystem processes (e.g., regulation of climate or floods); non-material benefits (e.g., recreational or aesthetic); and services necessary for the production of all other services (e.g., soil formation or nutrient cycle) (Decree-Law no. 142/2008, July 24th).

Educational community – in Portugal, they are composed of individual schools and school clusters; it includes the community of teachers and students, school administration and management bodies, operational assistants, psychology service and social action services, parents and parent associations.

Entomofauna – community of insects of an environment or region.

EU Pollinator Initiative – the European framework to tackle wild pollinator decline; the initiative is part of the EU biodiversity strategy and sets strategic objectives and actions to be taken by the EU and its Member States to address pollinators decline and contribute to global conservation efforts [31].

Farmers and forest leaders – emblematic farmers or forest managers that passionately believe in their vision and are seen as pathfinders, inspiring others to follow them.

Farmers organizations – collective persons with a non-profit associative base that group together farmers or other agricultural professionals (in the broadest sense, including agriculture, livestock, beekeeping and forestry) engaged in the development of activities in the agricultural field; farmers organizations contribute to the fulfilment of individual needs felt by their associates, including representation, defence and promotion of their interests.

Forests – refers to land with forestry use, occupied by forest trees, or temporarily deforested because of cultural cuts; include trees originated by natural regeneration, sowing or planting with height ≥ 5 m and coverage ≥ 10% [30].

Forest intensification – forest system where management practices aim to increase the yield per unit of forest.

Forest organizations – also known as forest owner's associations or woodland owner's groups, are organized groups or associations that represent and advocate for the interests and concerns of private forest landowners; these organizations are typically formed by individuals or entities who own and/or manage forested lands for various purposes.

Formal education – the hierarchically structured, chronologically graded education system, running from primary school to the university and including, in addition to general academic studies, a variety of specialised programmes and institutions for full-time technical and professional training [32].

Governmental Conservation Organizations – refers to all governmental organizations involved with nature conservation in Portugal mainland and Autonomous Regions.

Green and blue infrastructures – defined as a strategically planned network of natural and semi-natural areas [green (land) and blue (water) spaces] with other environmental features, designed and managed to deliver ecosystem services and enhance biodiversity.

Habitats Directive – Council Directive 92/43/EEC comprising 24 articles of legislation adopted by the European Union in 1992 as a response to the Berne Convention, to which all the member states must comply, to promote the conservation of natural habitats and of wild fauna and flora [33].

Honeybee colony – beekeeping production unit, consisting of the physical support (usually a beehive), the bee swarm, the brood and all biological materials produced by the bees.

Informal education – the truly lifelong process whereby every individual acquires attitudes, values, skills and knowledge from daily experience and the educative influences and resources in his/her environment [32].

Insect-pollinated plants – plants that rely on insects (e.g., bees, butterflies, hoverflies) to transfer the pollen grains from the anthers (or other specialized structures of the flower) of one flower to the stigma of the same or of another flower of the same species.

Invasive species – non-indigenous species whose introduction into nature or propagation in a given territory threatens or has an adverse impact on biodiversity and ecosystem services associated with it (Decree-Law no. 92/2019, July 10th).

Key actors – refers to all the actors that may be, directly or indirectly, involved in pollinator conservation and promotion actions; examples are local and central administration; beekeepers, farmers, forest managers and their organizations; municipality technicians, companies that manage urban green spaces and landscape architects; academia, including researchers and research organizations; non-governmental organizations; educational and elderly communities; and citizens.

Key dissemination actors – refers to the actors involved in dissemination activities, including Museum and Science Centres, municipalities, non-governmental organizations, and farmers, forest and beekeepers' organizations, farmers and forest leaders, local and central administration, as well as actors from academia, media, companies, and collaborative networks (e.g., polli.NET).

Landscape – the territory, as perceived by the populations, whose character results from the action and interaction of natural and/or human factors (adapted from the definition established by the European Landscape Convention, Decree-Law no. 4/2005, February 14th).

Local administration – local and regional governmental divisions which includes Municipalities and Parish Councils.

Managed pollinators – a pollinator that is maintained by human beings through husbandry (e.g., honeybees, bumblebees) [34].

Management practices – working methods and innovations used to improve the effectiveness of work systems and to optimize resource use.

Minimum viable scheme – refers to part of the European Pollinator Monitoring Scheme (EU-PoMS) that proposes a standardised methodology to be implemented by Member states to monitor species abundance and diversity, and occupancy data on wild bees, butterflies and hoverflies [35].

Monitoring – process of periodically collect, analyse, and use information to actively and effectively manage performance, maximising positive impacts and minimising the risk of adverse impacts.

National Action Plan for Pollinators – plan for action defined at the national level that identifies pollinator conservation actions, key actors, and funds for action implementation, and includes the implementation of a pollinator monitoring scheme.

National Pollinator Strategy – general plan defined at the national level that is based on strategic thinking and strategic planning to achieve long-term overall goals under conditions of uncertainty; it establishes priorities, identifies work axes, objectives, and measures for pollinator conservation, and identifies the need for a National Action Plan for Pollinators.

National Register of Classified Natural Values (*Cadastro Nacional dos Valores Naturais Classificados*) – archive listing the classified natural values and plant or animal species that are assigned to a threat category by the national authorities in accordance with international criteria defined by International Union for Conservation of Nature (Decree-Law no. 142 /2008, in its current version).

National Strategy for the Conservation of Nature and Biodiversity 2030 – instrument for the pursuit of environmental policy and a response to national and international responsibilities to reduce the loss of biodiversity, having been published by the Resolution of the Council of Ministers n.° 55/2018 - *Diário da República* n.° 87/2018, Series I; the ENCNB 2030 highlights three strategic points: improving the state of conservation of the natural heritage, promoting recognition of the value of the natural heritage, and fostering the appropriation of natural values and biodiversity by society.

Native species – any species originating from a certain territory or having a natural area of distribution there, past or present, excluding its hybrids with exotic species (Decree-Law no. 92/2019, July 10th).

Natural and semi-natural areas – areas with spontaneous or sub-spontaneous vegetation [36].

Non-formal education – any organised educational activity outside the established formal system – whether operating separately or as an important feature of some broader activity – that is intended to serve identifiable learning audience and learning objectives [32].

Pastures – refers to areas with or without human intervention occupied essentially with herbaceous vegetation, whether cultivated (sown) or natural (spontaneous), not included in a farm rotation system.

Pollination – refers to the transfer of pollen grains from the anthers (or other specialized structures of the flower) to the stigma of the same or of another flower of the same species; this process allows fertilization and the production of fruits and seeds in wild plants and crops.

Pollinator – here it refers to any animal that systematically carries the pollen grains from the male structures of the flower to the female structures of the same or of another flower of the same species, thus, directly contributing to pollination; it includes, for example, bees, flies, beetles, butterflies, bats and birds, rodents and lizards.

Pressures – factors that negatively affect ecosystems and their biodiversity.

Producer's organizations – refers to farmer, forest and beekeeper's organizations.

Protected areas – areas identified by the National System of Classified Areas (NSCA) defined by Decree-Law no. 142/2008, July 24th, and Decree-Law no. 242/2015, October 15th, including the areas of the National Network of Protected Areas (RNAP), Classified Areas that are part of the Natura 2000 Network and other Areas classified under international commitments assumed by Portugal [37].

Researchers – individuals and organizations that belong to the National Scientific and Technological System (NSTS, *Sistema Científico e Tecnológico Nacional*) as defined by The Foundation for Science and Technology, I. P. (FCT, *Fundação para a Ciência e a Tecnologia*).

Science communicators – Anyone (person and organization) who communicates to increase the impact of science in society.

Shrubland – refers to natural areas of spontaneous vegetation, little or very dense, in which the shrub cover (e.g., heather, brambles, broom, gorse, wild boar) is ≥ 25%; it also includes abandoned olive groves if less than 45 trees/ha [30].

Theory of Change – method that identifies the path (process) that can be followed so that a certain result (change) is achieved or occurs, through the sharing and discussion of ideas between various actors; this method produces a narrative in the form of a diagram that links activities and results in a cause-effect logic.

Urban areas – refers to artificialized areas intended for human intervention activities; it includes built areas, industrial areas, commercial areas, areas dedicated to tourism, infrastructures, road and railway networks, service areas, gardens and equipment [30].

Wild plants – plants that grow alone in natural ecosystem or semi-natural system and remain alive without the help of direct human interaction.

Wild pollinators – native animal species that facilitate pollination of crops and wild plants and survive in the wild without the aid of human beings through husbandry.

BIBLIOGRAPHIC REFERENCES

[1] Resolução da Assembleia da República n.º 336/2021, Diário da República, 1ª série, n.º 245, 21 de dezembro de 2021.

[2] Resolução do Conselho de Ministros n.º 55/2018, Diário da República, 1ª série, n.º 87, 7 de maio de 2018, Estratégia Nacional de Conservação da Natureza e Biodiversidade 2030.

[3] Borges PAV, Lamelas-Lopez L, Andrade R, Lhoumeau S, Vieira V, Soares AO, Borges I, Boieiro M, Cardoso P, Crespo LCF, Karsholt O, Schülke M, Serrano ARM, Quartau JA, Assing V. 2022. An updated checklist of Azorean arthropods (Arthropoda). *Biodiversity Data Journal* 10: e97682. https://doi.org/10.3897/BDJ.10.e97682

[4] Kratochwil A, Smit J, Aguiar AF. 2018. Updated checklist of the wild bees of the Madeira Archipelago and the Selvagens Islands (Hymenoptera: Apoidea: Anthophila). Linzer Biologische Beiträge 50: 1213-1228.

[5] Portal da Biodiversidade dos Açores, https://azoresbioportal.uac.pt/pt/

[6] Browning PRG. 2019. Butterflies of the Iberian Peninsula - A photographic record and field guide.

[9] Wood T, Ortiz-Sanchez FJ. 2022. Description of three new *Andrena* Fabricius, 1775 species from understudied parts of Iberia (Hymenoptera: Andrenidae). Boletín de la Sociedad Entomológica Aragonesa 70: 114-123.

[10] Gaspar H, Wood T, Siopa C, Lopes S, Loureiro J, Castro S. 2022. New regional contributions to the knowledge of the portuguese bee fauna (Hymenoptera: Anthophila). Boletín de la Sociedad Entomológica Aragonesa 71: 54-62.

[11] Cross IC. 2023. A new species of *Hoplitis* (*Stenosmia*) from southern Iberia (Hymenoptera: Megachilidae). Entomologist's Monthly Magazine 159: 79-86.

[12] van Eck APW, Andrade R, van Steenis W, van der Ent J. 2020. New additions to the hoverflies of mainland Portugal (Diptera, Syrphidae) with some observations on flower visits. Boletín de la Sociedad Entomológica Aragonesa 66: 193-198.

[13] Habitats Directive, Council Directive 92/43/EEC, 21 May 1992, conservation of natural habitats and of wild fauna and flora.

[14] European Butterfly Monitoring Scheme Portugal - Censos de Borboletas de Portugal, https://butterfly-monitoring.net/pt-pt/portugal-bms

[15] Rede de Estações de Borboletas Nocturnas, https://www.reborboletasn.org/

[16] SPRING: Strengthening Pollinator Recovery through INdicators and monitorinG, https://pollinator-monitoring.net/

[17] Ollerton J, Winfree R, Tarrant S. 2011. How many flowering plants are pollinated by animals? Oikos 120: 321-326.

[18] Carta de uso e ocupação do solo (COS) para 2018, DGT - Direção-Geral do Território.

[19] Instituto Nacional de Estatística (Statistics Portugal), https://www.ine.pt

[20] Potts SG, Imperatriz-Fonseca V, Ngo HT, Aizen MA, Biesmeijer JC, Breeze TD, Dicks LV, Garibaldi LA, Hill R, Settele J, Vanbergen AJ. 2016. Safeguarding pollinators and their values to human well-being. Nature 540: 220-229.

[21] Borges PAV, Abreu C, Aguiar AMF, Carvalho P, Jardim R, Melo I, et al. 2008. A list of the terrestrial fungi, flora and fauna of Madeira and Selvagens archipelagos. Funchal and Angra do Herísmo: Direcção Regional do Ambiente da Madeira and Universidade dos Acores.

[22] Borges PAV, Costa A, Cunha R, Gabriel R, Gonçalves V, Martins AF, et al. 2010. A list of the terrestrial and marine biota from the Azores. Princípia (ISBN: 978-989-8131-75-1).

[23] Weissmann JA, Picanço A, Borges PA, Schaefer H. 2017. Bees of the Azores: an annotated checklist (Apidae, Hymenoptera). ZooKeys 642: 63.

[24] Kratochwill A, Schwabe A. 2018. Wild bees (Anthophila) of Macaronesia: Aspects of biogeography, species diversity, evolutionary patterns and taxonomic problems. Mitteilungen des Entomologischen Vereins Stuttgart 53: 10-12.

[25] Rego C, Smit J, Aguiar AF, Cravo D, Penado A, Boieiro M. 2022. A pictorial key for identification of the hoverflies (Diptera: Syrphidae) of the Madeira Archipelago. Biodiversity Data Journal 10.

[26] Esposito F, Costa R, Boieiro M. 2021. Foraging hehavior and pollen transport by flower Visitors of the Madeira Island endemic *Echium candicans*. Insects 12: 488.

[27] Picanço A, Rigal F, Matthews TJ, Cardoso P, Borges PAV. 2017. Impact of land-use change on flower-visiting insect communities on an oceanic island. Insect Conservation and Diversity 10: 211-223.

[28] Boieiro M, Ceia H, Caramujo MJ, Cardoso P, Garcia-Pereira P, Pires D, Reis J, Rego C. 2023. Livro Vermelho dos Invertebrados de Portugal Continental. FCiências.ID e ICNF I.P., Lisboa. https://lvinvertebrados.uac.pt/docs/Livro-Vermelho-dos-Invertebrados-de-Portugal-Continental.pdf

[29] polli.NET – Rede colaborativa para a Avaliação, Conservação e Valorização dos Polinizadores e da Polinização: https://www.pollinet.pt/

[30] Carta do Uso e Ocupação do Solo 2018, National Territory Department – Direção Geral do Território: https://www.dgterritorio.gov.pt.

[31] European Pollinator Initiative: https://environment.ec.europa.eu/topics/nature-and-biodiversity/pollinators_en

[32] Smith MK. 2002. Informal, non-formal and formal education: a brief overview of different approaches. The encyclopaedia of pedagogy and informal education. https://infed.org/mobi/informal-non-formal-and-formal-education-a-brief-overview-of-some-different-approaches/

[33] Habitat Directive: https://environment.ec.europa.eu/topics/nature-and-biodiversity/habitats-directive_en

[34] IPBES. 2016. The assessment report of the Intergovernmental Science-Policy Platform on Biodiversity and Ecosystem Services on pollinators, pollination and food production. SG Potts, VL Imperatriz-Fonseca, HT Ngo (eds). Secretariat of the Intergovernmental Science-Policy Platform on Biodiversity and Ecosystem Services, Bonn, Germany. https://doi.org/10.5281/zenodo.3402856

[35] Potts S, Dauber J, Hochkirch A, Oteman B, Roy D, Ahnre K, Biesmeijer K, Breeze T, Carvell C, Ferreira C, Fitzpatrick Ú, Isaac N, Kuussaari M, Ljubomirov T, Maes J, Ngo H, Pardo A, Polce C, Quaranta M, Settele J, Sorg M, Stefanescu C, Vujic A. 2020. Proposal for an EU Pollinator Monitoring Scheme. Publications Office of the European Union, Luxembourg. https://publications.jrc.ec.europa.eu/repository/handle/JRC122225

[36] Capelo J, Mesquita S, Costa JC, Arsénio P, Neto C, Monteiro-Henriques T, Aguiar C, Honrado J, Espírito-Santo MD, Lousã M. 2007. A methodological approach to potential vegetation modeling using GIS techniques and phytosociological expert-knowledge: application to mainland Portugal. Phytocoenologia 37: 399–415.

[37] ICNF – Instituto da Conservação e das Florestas, Protected areas: https://www.icnf.pt/oquefazemos/materiaisinformativoseeducativos/areasprotegidas

APPENDIX 1

Team gathered for the development of the Portuguese Pollinator ToC and respective institutions, institution typology and main field of action. *Country coordinators, ♦Consultants.

Name	Institution - Name	Institution typology	Main field of action
Sílvia Castro*	Promote Pollinator Contact point & polli.NET & CFE - Centre for Functional Ecology, University of Coimbra	National and International Networks, Academy	Biodiversity & Conservation
João Loureiro*	CFE - Centre for Functional Ecology, University of Coimbra	Academy	Biodiversity, Ecology & Evolution
Marten Schoonman♦	Naturalis Biodiversity Center	National and International Networks, Academy	Biodiversity & Conservation
Martijn Thijssen♦	Promote Pollinators	Non-Governmental Organization, International Networks	Policy, Biodiversity & Conservation
Anabela Nave	INIAV - Instituto Nacional de Investigação Agrária e Veterinária	Public administration	Agriculture & beekeeping
Andreia Penado	Ciência Viva – Agência Nacional para a Cultura Científica e Tecnológica	Education & Science dissemination	Science communication, Insect diversity
Adriana Galveias	Instituto Politécnico de Setúbal	Education & Science dissemination	Science communication
Bárbara Oliveira	DGAV - Direção Geral da Alimentação e Veterinária	Public administration	Agriculture & beekeeping
Carla Rego	SPEN - Sociedade Portuguesa de Entomologia	Non-Governmental Organization, Academy	Insect diversity
Cláudia Fernandes	Department of Geosciences, Environment and Spatial Planning, School of Sciences, University of Porto	Academy	Landscape Architecture
Elisabete Figueiredo	ISA - Instituto Superior de Agronomia, Universidade de Lisboa LEAF - Linking Landscape, Environment, Agriculture and Food	Academy	Agriculture & Ecology
Eva Monteiro	Tagis - Centro de Conservação das Borboletas de Portugal	Non-Governmental Organization	Nature Conservation
Helena Ceia	ICNF - Instituto da Conservação da Natureza e das Florestas	Public administration - Conservation	Nature Conservation

Name	Institution - Name	Institution typology	Main field of action
Helena Freitas	IPBES Contact point & CFE - Centre for Functional Ecology, University of Coimbra	International organizations, Academy	Biodiversity & Ecosystem functioning
Inês Santos	Syngenta	Industry	Agrochemicals
Joana Godinho	INIAV - Instituto Nacional de Investigação Agrária e Veterinária	Public administration	Agriculture & beekeeping
João T. Nunes	ICNF - Instituto da Conservação da Natureza e das Florestas	Public administration	Nature Conservation
Luís Guilherme Sousa	Câmara Municipal da Lousada	Public administration, Municipality	Biodiversity & Conservation
Maria Luís Fino	CAP - Confederação de Agricultores de Portugal	Producers Association	Agriculture
Mário Boieiro	Grupo da Biodiversidade dos Açores, Universidade dos Açores	Academy	Biodiversity & Conservation
Mário Carmo	Mário Carmo ~ Your Biodiversity & Ecosystem Consultant	Industry	Environmental consulting
Ofélia Anjos	IPCB – Instituto Politécnico de Castelo Branco Polytechnic University CERNAS-IPCB - Research Centre for Natural Resources, Environment and Society	Academy	Beekeeping
Sandra Antunes	NBI - Natural Business Intelligence	Industry	Environmental consulting
Sofia Ferreira	FSC Portugal – Associação para uma Gestão Florestal Responsável	Non-Governmental Organization	Forest Certification
Sónia Ferreira	Associação BIOPOLIS / CIBIO - Centro de Investigação em Biodiversidade e Recursos Genéticos da Universidade do Porto	Academy	Biodiversity

IMAGE INDEX

Eristalinus taeniops Wiedemann, 1818 (Syrphidae, Diptera) pollinating the flowers of *Oenanthe crocata* L. (Apiaceae)
João Loureiro 6

Andrena sp. (Andrenidae, Hymenoptera) visiting the flowers of *Coleostephus myconis* (L.) Rchb.f. (Asteraceae)
João Loureiro 13

Zerynthia rumina Linnaeus, 1758 (Papilionidae, Lepidoptera) in *Lavandula* sp. (Lamiaceae)
Sílvia Castro 14

Myathropa florea Linnaeus, 1758 (Syrphidae, Diptera) pollinating the flowers of *Aster amellus* L. (Asteraceae)
Sílvia Castro 18

Lycaena phlaeas Linnaeus, 1761 (Lycaenidae, Lepidoptera) in *Lavandula pedunculata* (Mill.) Cav. (Lamiaceae)
Sílvia Castro 20

Portuguese landscape illustration
Margarida Moreira 22-23

ToC diagram
Todos os autores 32-33

Eristalis tenax Linnaeus, 1758 (Syrphidae, Diptera) pollinating almond flowers (*Prunus amygdalus* Batsch, Rosaceae)
Sílvia Castro 36

Andrena albopunctata Rossi, 1792 (Andrenidae, Hymenoptera) pollinating sunflower (*Helianthus annuus* L., Asteraceae)
Sílvia Castro 51

Omphalodes nitida Hoffmanns. & Link, Boraginaceae), a plant pollinated by insects with very slender proboscises
Sílvia Castro 52

Shrubland with *Lavandula pedunculata* (Mill.) Cav. (Lamiaceae), a plant whose flowers are highly attractive to numerous insect pollinators
Sílvia Castro 54

Bombus vestalis Geoffroy, 1785 (Apidae, Hymenoptera) visiting the flowers of *Arbutus unedo* L. (Ericaceae)
João Loureiro 56

Anthophora plumipes Pallas, 1772 (Apidae, Hymenoptera) observing the flowers of *Gladiolus communis* L. (Iridaceae)
Sílvia Castro 58

Melitaea deione Geyer, 1832 (Nymphalidae, Lepidoptera) pollinating the flowers of *Oenanthe crocata* L. (Apiaceae)
João Loureiro 66

www.ingramcontent.com/pod-product-compliance
Lightning Source LLC
LaVergne TN
LVHW022027200926
848481LV00007B/2569
9789892626260